Railing Installation Profitable Trade Secrets

By Butch Berardesco
(a.k.a. Butchie9Fingers)

TheRailingAuthority.com

About the Author

Butch has over 30 years experience in the railing installation business without ever missing a C/O date. His diversified background in everything from working on an assembly line to building custom motorcycles to land surveying, has given him an edge in layouts, custom tool design and production. His skill has taken him from Hawaii to Anguilla and everywhere in —between. He attributes his success to prayer and faith in God whom he says always makes him look better than he is.

Business Philosophy:

Doing what is important so that nothing becomes urgent, thus eliminating working nights and weekends.

My favorite Murphy's Law, is "Before you can do something, you have to do something else first." That means get that done, before the railing is delivered.

Develop a plan before the railing arrives, break it down into phases and share your plan with the G/C. Advise him well in advance of what work needs to be done by other trades, before you can start the next phase. Keeping him well—informed of what is important will reduce the urgency and will eliminate any "down—time" for you.

Most importantly, you run the job; do not let the job run you. Develop the mindset that you are there to complete the next phase, not merely to go to work.

Table of Contents

Chapter 1
Meeting with Contractors

Determine anticipated c/o date at first meeting with G/C.

Discuss a plan of action, and which of the five installation options would be suitable.

Option A: install railing floor by floor, from the top down, after stucco is complete, balcony edges are cleaned and exterior painting is complete.

Advantages

1. Minimal punch out

2. Minimal downtime, no waiting on other trades.

3. Up and down transportation is minimizes,

4. Installation can be complete in a very timely manner

Option B: install railing in drops, from the top down, north side, east side, south and west sides.

Follow behind stucco and painting, and after the building is topped off and the stucco is cleaned off the balcony edges.

Advantages

1. The building can be stocked more efficiently in drops

2. The railing can be built more efficiently in drops

Disadvantages

1. Downtime — waiting for drops to be released

2. Downtime — waiting for transportation

3. Backtracking to each floor, several times

Option C: installing railing from the bottom up before the building is topped off. After stucco and painting and covering the finish product with plastic.

Disadvantages

1. Considerable downtime – do to concrete – pouring concrete has top priority, which is understandable. It results in increased ground traffic. Slowing down, even stopping the offloading of deliveries.

2. The concrete process uses almost all available ground space with steel, forms, and equipment. Finding a safe convenient, storage or staging area extremely difficult

3. All work stops where form tables, are being moved.

4. Concrete uses 95% of the crane time

5. Concrete blowouts do occur, putting the finished product in jeopardy

6. Covering the railing adds an additional very time consuming process

7. The installation process can take as much as 3-5 times longer

Steps to take if "C" is the only option

1. A designated storage is a must

2. Must have priority when it comes to scheduled equipment time for unloading, moving and stocking the floors including forklift, buck hoist, elevators, and/or crane

Things the G/C must be aware of

1. Railing is a finished product and should never take the place of safety rail

2. Covering railing with plastic only protects against paint splatter and light stucco.

3. Plastic has a relatively short life span and is adversely affected by sun, rain, and especially wind plastic covering will have to be maintained on a continual basis.

4. Contractor must make thin plywood cap covers over a soft buffer material to protect railing against unfinished sheer walls from overhead chipping, concrete spills, and heavy stucco.

Wood covers should be made in standard lengths and moved around the building as necessary.

5. Tar-paper should be slid under railing bottom channel to protect the slab edge against heavy stucco. This will minimize or eliminate grinding after railing installation

Option D: installing railing from bottom to top, one floor behind the block trade, before stucco or painting

Advantages: to manufacturer

1. Railing would be installed, punched out, signed off and covered with plastic.

2. No completion push to manufacturer

Disadvantages: to G/C

1. Same general conditions as option "C" with less ground space and damage control.

2. Once the railing is signed off the G/C is responsible.

3. Continual maintenance keeping railing protected.

Tips to G/C

1. Use plywood covers next to sheer walls

2. Cover balcony edges with tarpaper before stucco

Option E: installing from a swing stage:

Remodeling jobs with no inside access;

1. Contractor should supply and maintain swing stages as other trades will be using them

2. Swing stages must be moved in a timely manor

3. The more swing stages, the faster the job will go

4. At least two swing stages are required for railing

5. Slab painting should be done after installation, or at least after core drilling

6. Residents may not have balcony access until new railing is installed No same day access!

Advantage

With proper co-operation between the contractor and other trades, this method can be a very expedient way to install railing with little or no punch out.

Option F: The preferred method.

Advantage

1. No storage

2. No double handling

3. No damage or protecting.

4, No downtime

5. Minimal punch-out

Most G/C's are under the misconception that railing installation is a long drawn—out process, so they want the railing on the jobsite ASAP. However, when the railing arrives on the jobsite prematurely, it quickly becomes a long, drawn—out, chaotic process.

Under the right conditions, I completed an 11—story building in 6 days, after the railing arrived. I also completed a 26 story building in just 35 days, after the railing arrived. I could give a long list but each job has a single commonality— After the railing arrived. What I did not mention is the preparation time needed to check every hole with templates, make sure every hole is in the right place and deep enough and that the steel is cut out and painted before the railing arrived. All the chipping and core drilling is completed before stucco or painting; eliminating the mess factor. All the parapet walls are core—drilled, cores popped and the holes filled with sand. Stucco trade stucco's right over the sand without interference.

With the proper advance slab preparation, the railing can then be delivered directly into the staging area, after the exterior of the building is complete. While in the staging area any railing preparation, like attaching inter splices or repairs, can be made in a single location — production style. From there, the railing can be lifted, connected and placed safely into the holes to be poured.

Balconies can be completed from the top down, in drops, or floor—by—floor, depending on the stocking system used.

Committing to the G/C's anticipated completion date

The G/C will want a commitment from you to meet their anticipated c/o. This is the time to establish the following necessary criteria that the G/C must meet, before you can commit to their anticipated c/o date; especially if he chooses any other option other than Option F: The Preferred Method.

1. There must be an assigned storage and staging area for the railing, that is safe and convenient.

2. Storage area may need to be re—assigned as portions of the building are completed.

3. If the next portion of the building is not ready for railing installation as scheduled, the manufacture will hold the shipment until the building is ready. However, product payment will still be made by G/C. G/C will provide an additional storage area and assume full responsibility for any damage and additional handling to move the railing to designated storage area.

4. Equipment used to unload or move railing via forklift, crane, buck hoist, or elevator, will be scheduled in advance and will have priority after concrete and stucco, but before dumpster trash, and other trades. If another trade did not schedule enough time for a piece of equipment, they will relinquish control

until our scheduled time has expired.

5. G/C is responsible for transportation via buck hoist and elevators. Both must be run with maximum efficiency.

Trades must have priority over trash. As elevators are brought on line, they must be protected from damage and staffed immediately. Hoist and elevators must have radios.

6. G/C must supply convenient water and power. A two—hour warning must be issued if/when either is to be turned off.

7. Units must be left unlocked or the railing installer must have a key. The key is NOT negotiable, even if it must be signed out daily.

Chapter 2
Handling Deliveries

A. Unloading Trucks, And Moving Material - Plain Of Action

Handling deliveries is one of the most important parts of the job, there is no best way, but there is always a better way.

Schedule delivers, well in advance, and be prepared.

Plan every detail. The area where the railing sits, to be lifted to the balconies is called the staging area.

Bring the railing as close to the staging area as possible. Plan exactly where the truck backs in. How it is to be unloading, via forklift, crane or by hand. Know exactly where the railing is going and what direction it must face.

B. Labor

Quick labor for unloading, moving material, stocking, or installation, can be obtained through labor staffing companies.

Open an account with a few nationwide companies in advance, so labor is readily available. Once an account is opened locally, it can be expanded to another city or state by phone, the day before.

Order labor the day before deliveries.

The advantage of using labor staffing companies is worker compensation, and employee problems, are not an issue. In addition, if a regular is a no—show, a replacement is sent automatically. You are never shorthanded.

The G/C usually has day laborers. It is possible to negotiate some of his help for short periods, even after hours, for a minimal amount of out— of— pocket cash paid directly to the laborer. Take full advantage of the exemplary laborers by offering them a cash bonus, it will bring out their best.

C. Moving Material By Hand

This is done with two people, moving two pieces at one time. One person at each end, standing between the two railings. They simultaneously lift the railing, one piece in each hand and walk. Two people can carry two pieces long distances with a minimum amount of effort.

One person caring one piece is very strenuous because of length, and balance.

D. Unloading With A Forklift

Find out where the contractor rented his forklift. Arrange with the rental company to have you and your crew certified. This only needs to be done once. The certification will act as a license so laminate it and keep it with you. Without it the contractor will refuse to let you, or someone in your crew operate the forklift. If possible, do not be dependent on someone else's driver!

Equipment certifications are changing continuously; you may be forced to use their designated operator. Make friends with him; buy him an occasional snack or a drink off the lunch truck. A little generosity will go a long way. Negotiate the use of the equipment, at the first meeting. Schedule all equipment needs, well in advance. Larger jobs will have a ground control man. He is **very important.** Again, "make friends."

If the bottom channel of the railing is going directly on the steel forks, have designated carpet or wood covers that easily attach to the top of the forks, to keep the railing from sliding.

(Unloading with a crane; see next chapter)

Chapter 3
Stocking the Building

A. Manual Lifting

Lifting up one floor at a time:

Tie off, then lift one end of the railing to a man on the floor above; when he has it, walk to the other end and lift it to another worker, they lift together till it clears the slab and pull it in. Do not drag it across the slab edge before it clears, as this will cause damage to the railing and stucco. You may need to bend cardboard over the edge for protection.

If there is only one man on the floor above, lift one end till he has it, walk to other end and as you are lifting it up to him, he is working his way to the center of the rail, and pulls it in after he clears the slab.

If extra laborers are available, place one man on every floor or every other floor (if the railing is long enough) and send it up like a ladder. A man can be on the roof with a rope clipped to the top end of the rail to brace and stabilize it as it goes up.

This could be done after hours with the contractor's laborers, for a minimal amount of out—of—pocket expense.

B. Buck Hoist

Measure the length of the longest piece of railing, to see if the longest piece can be carried from the hoist, through the hallway, and into the rooms and out on the balcony.

The best way to determine this is to slide two metal studs together the length of the railing and do a dry run.

Now, measure the hoist; length, width, height, diagonally – corner to corner, and diagonally — ceiling to floor; every inch will be important.

Load all pieces that fit in hoist and can be carried down hallways, onto balconies.

Some pieces can be maneuvered down hallways and onto balconies, yet are too long for hoist. You can open the rear hoist door by disconnecting the limit switch on the hoist door.

The hoist will ride with the door open.

Law requires the contractor to build a very strong canopy roof at rear of hoist, to protect workers from falling objects while waiting for hoist.

Many times railing can be unloaded directly from truck onto canopy and into hoist. The railing will hang out the back of hoist, so attach safety rope inside the hoist 2 — 3 feet in from the open door, above the railing. This is to prevent someone from carelessly walking backward out the open door. This also allows you to safely unload, without being tied off.

Delivery and hoist times should be scheduled for just after quitting time, no one will be in your way and it will go quickly.

An inexperienced contractor may not agree, but assure him that, this is done on every job. Be sure to have your own Allen wrenches on you to disconnect the limit switch.

If the railing doesn't fit through the room to the balcony where it's to be installed, but fits into a nearby room and balcony, the railing may be passed from one balcony to the next. If there is a large gap between balconies, the railing can be swung by a rope with spreaders to the railing from the roof, and a tag line from each attached from the spreaders to each balcony. (Spreaders and taglines to be explained in detail in the following chapters.)

C. Manual Hoist: ropes & pulleys

Lifting rail by hand using ropes and pulleys, can be used on smaller buildings, with as little as three people.

Purchase two well pulleys, the kind that stucco laborers use.

You also need two out riggers; you can rent them or have them custom made to fit over roof parapet wall. They should be adjustable and extend out two feet. In a pinch, you can nail three 2x4's together, tie rope 5/8" or bigger around one end to hang a pulley from. Slide the other end into an old pallet and add about six or more buckets of grout for weight. Repeat this process and make another outrigger.

Place them approximately eight feet apart, on the top floor balcony and hang the pulleys off them.

Use at least 1/2" diameter rope, poly—propylene is relatively inexpensive in a 600' spool. Thread the rope through pulley. Attach a large spring link hook to one end of rope. Attach another hook approximately 18" to 2' up from the first.

The first hook is passed under the cap and clipped to the second hook without touching the rail as it rides.

Tie the other end of the rope just above the second hook. This will act as a tag line on the way up, and a way to pull the hooks down; it will also form a circle. Clearly mark the end you are pulling with tape or paint. Pulling the wrong end may be hazardous.

There should be two out riggers, two pulleys, two ropes, four large spring link hooks for quick connect and disconnect, and two people pulling up one piece of rail.

Working as a team is very important. Instruct those lifting that the low end is heavier. A stronger man will have a tendency to lift faster which will cause his end to become higher — adding more weight to the other end. Keep the rail as even as possible going up.

If the railing is too heavy for two people, you can do the following:

1 Two more men can be added to team with first two men on ground, four hands on each rope working in unison.

2. A third man, rope & pulley can be added.

3. Hook up the railing to be lifted, pull the slack out of the rope, have the man on balcony loop a knot and add a hook on the pull rope side. Then rope one or two gallon jugs of water to the hook. The load can be lightened by 8 lbs. per gallon. (if water jugs fall to ground and break open, nothing on ground is damaged).

D. Cable Operated Winches (electric powered)

Electric powered cable operated winches are swing stage motors that have been set up to work upside down.

They hang from a swing stage outrigger, clamped either to the parapet wall around the roof, or to a weighted down outrigger attached to a scaffold on the roof. They are operated by a wire remote control from the roof. The cable feeds through the motor and collects in a pile on the roof.

Advantages

1. They are easy to set up and not too difficult to move.

2. They can be rented everywhere.

3. No cable limit, no height limit.

4. Operator can see clearly from the roof.

Disadvantages:

1. Very slow, thirtyfive feet pre—minute up, thirtyfive feet per—minute down.

E. Cable Operated Winches (gas powered)

Gas powered winches are usually set up on the ground, with 800 to 1000 lbs. of weight holding it down. The cable travels from the winch up through a pulley mounted to a swing stage parapet outrigger clamp, a weighted down outrigger on the roof or a outrigger braced to the balcony floor or ceiling, with screw jacks. Then the cable travels back down to the load.

Advantages

1. Much faster, seventy to ninety feet per minute

2. No dependency on the electrician to wire in a 220 plug & move it if need be.

Disadvantages

1. The cable is on a drum and must be watched closely. It may need help to feed properly.

2. Very loud, you must depend on visual only.

3. Height is limited due to cable drum

However, there is a way to increase the height. If you set the winch up on a balcony, and the outrigger on the floor above, the cable travels through the pulley on the outrigger above to the load on the ground.

Caution

The winch has to be set up perfectly square with the edge of the balcony, with the cable to the outrigger in perfect alignment with the cable from the outrigger.

If not, the cable will feed to one side of the drum. Use 1000 lbs. of weight to hold the winch down. The winch will be very close to the edge, so a ground visual can be maintained while the winch is being operated.

Caution

Operator must stay alert! If one end of the railing is about to hit the building, stop immediately and wait until railing stabilizes before continuing. If the railing gets hung up under a balcony slab and you don't stop, the winch with 1000 lbs. of weight will be picked up enough to go over the side. It only takes an inch or two. In addition, the winches have steel bottoms. Attach wood to the steel bottom. Sometimes they leak oil, you do not want oil between the slab, and the steel bottom because it will slide. Wood will prevent this. This is a very dangerous method and should only be considered as the last resort.

F. Cranes: Lifting Railings

Always schedule crane time in advance. You may need to purchase radios, they can be costly. It is always a good idea to make friends with the general contractor and his supervisors, ask them in advance for their help, or to possibly borrow radios.

Learn **"crane language"** and hand signals.

All cranes can cable up, cable down, swing right, and swing left. (that is his right and his left, not yours!)

Gibb cranes — these are cranes with a rigged boom. It's boom up, boom down.

Telescopic boom cranes – it's boom up, boom down, boom in and boom out.

Tower cranes — trolley in, trolley out. (no boom) the crane operator has a lot going on, and they are known to be temperamental and unforgiving. Do not make mistakes!!!!

Using a crane to stock the floors

Buy your own crane straps— 20' long, 2" wide. Railing is seldom choked; meaning to loop the strap around the load and then slip one end through the loop on the other end, and hook that end to the crane spreader hook. The strap gets tighter as the crane lifts the load, hence the word "choked". This could damage the railing if you are lifting more than one piece, because it puts too much pressure on the caps. The railing will be cradled most of the time. Cradled means, to loop one end around the load and both ends of the strap on one spreader hook. Looping both ends of a 4" wide strap on one spreader hook is time consuming and difficult, that is why you use your own 2" wide straps.

A crane spreader is a steel ring with two or four chains, or cables the same length, attached to it.

Each chain or cable has a hook attached to it. The steel ring is placed over the cranes main hook. Spreaders allow the railing to be picked up horizontally. Each crane will have its own spreaders.

Lifting the railing:

First, determine the maximum amount of pieces to be lifted at one time, and how many pieces are to go on each balcony. Are there any radius pieces?

Only 4 or 5 straight pieces can be landed on a balcony by two men, at one time. After landing the pieces, they must lean the railing against a wall or other stable object.

Move the railing through the building to other balconies whenever possible. Allow the crane to lift to the same balcony as much as possible.

Pick a location on the ground to lift from, where the crane operator can see you. This way, one of the two people on the ground can give hand signals to the crane.

Place the railing in order of lift loads. Measure the width of the widest load; cut two 2"x 4", one or two inches longer than the widest load. If the picket space is less that 3 1/2" or the pickets from the different pieces do not align, rip the 2x4 down to 2x2.

Mark the center of the 2x2 and mark the center of the crane straps. Match up the centerlines and fasten the 2x2 to the crane straps using tie wire or duct tape. Do not use nails or screws; contractor might condemn your straps for safety reasons. (Don't ask how I know that!)

Place the straps through the railing with the 2x2 between the strap and the bottom of the cap. Space straps on the outside of the post, one section in. It does not matter if the railing has open ends— it will not slip out.

The wood under the cap keeps the pressure off the caps, the railings will ride straight and the bottoms stay together.

String can help hold the bottoms together if necessary.

The two workers receiving the load need a shepherds hook, made from electrical conduit or aluminum, to help pull the load in.

Because the load is cradled, they only have to disconnect one end of each strap off the spreader hooks, and the other end can stay on the spreader hooks at all times, while loading and unloading.

G. Lifting with a Crane Basket

This can be a great option if done right. It has tremendous advantages for jobs close enough to be delivered, by the manufacturer. No material stored on the ground, no forklift, winch or buck hoist.

Special baskets will have to be constructed as universal as possible. It must be able to be loaded at the shop and trucked to the jobsite. Most railing is less than twenty foot long. Road limits are eight foot wide in most states.

A good start would be 20' long by 8' wide and 4' high. It can be longer if the manufacturer has a truck or trailer that can handle it.

It can be designed to stack and ship two high, if the manufacturer is able to stack them at the shop.

They should be metal framed with wire mesh sides, a plywood bottom (so railing can be slid around while loading and unloading). The back can be left open with a strong safety cable clipped across the back for easy access.

Basket will need a quad spreader. Placement should be two cables 4' from the front, two cables 4' from the rear, so the cables will not hit the slab above when the crane is positioning the basket on the designated floor.

An 8' steel stiffener connecting the two rear cables, is necessary to keep the pressure off the rear gate of the basket while being lifted by the crane. The stiffener should be placed high enough to walk under while unloading. A front stiffener should be added for weight distribution.

Some jobs may require a basket constructed that can be unloaded from the side.

The railing must be loaded in a predetermined sequence or valuable crane time and more valuable the company's credibility will be lost. When everything is running smoothly, the contractor is more willing to assist, or at least leave you alone. If the contractor or crane operator feels you are inexperienced, everything you do will be scrutinized, and will then have to be approved. This means downtime and you will go to the back of the line and wait.

If done properly, the truck driver can hook up and hand signal the crane to lift. Then the whole crew, with no one on the ground, can unload the basket.

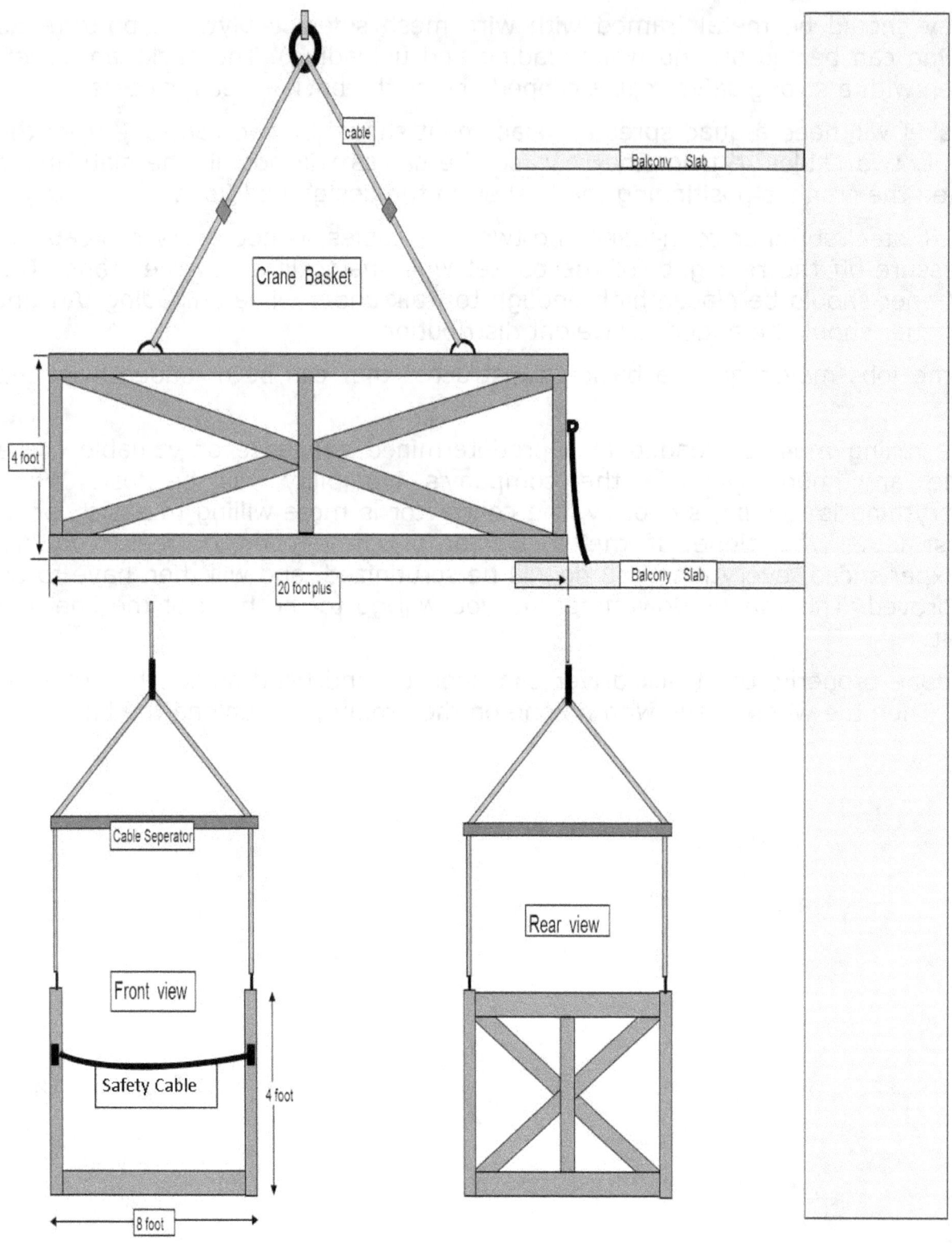

H. Stocking Building With Swing Stages

Swing stages can be as long as 55' with just two motors controlling it. They are easy to operate, assemble, and move.

Contact the rental company and become certified to assemble and move the swing stage. They will send a man out to the jobsite to conduct a class for you and your crew. This only has to be done once. Even if the contractor is responsible for moving the swing, you should know how to set it up to make sure, it is set up correctly. Your life does depend on it.

Swing stage should always be setup with the safety rail on the outside, so the front that faces the building will be lower.

If the railing fits inside the swing stage, be mindful of the weight. Most swing stages are rated for 1000 lbs. Each man is rated 250 lbs. No matter how much he actually weighs. That leaves 500 lbs. for railing. After the railing is loaded, use a pull strap or tie wire (coded solid copper wire, coded so it doesn't scratch the paint, solid so it can be twisted and doesn't have to be tied) to secure the railing to the back of the swing.

If the railing has a return that will protrude out past the front of the swing stage, use 2x4's across the width of the low part of the swing stage, not on top of the safety rail. Cut one 2x4 the size of the width outside to outside (or 27 1/2") and one the width inside to inside (23 1/2"). Center the 23 1/2 " piece on the 27 1/2 " piece, nail them together, and drop them in. The smaller piece will keep the larger piece from sliding off when the rail is being loaded and unloaded.

If the return is so long that it would hit the building, coming out of the front, remove a section of the rear safety rail. Then flip the rail over so the return over hangs the back, then slide the safety rail through the picket space and reconnect it. Another way is to make a safety rail out of rope with enough slack to accept the return, then tie wire it down to keep it in place.

It's always wise to rent wired remote controls that attach to the motors so you can operate the motors without reaching through the railing. Keep your full attention on what you are doing, keep the swing as level as possible, both motors may not run at the exact same speed, weight distribution is crucial. The most important thing is to keep the swing from getting caught on anything. At least one of the remote controls should have an extra—long cord so the swing could be operated by one man in case of an emergency.

I. Spreaders And Tag Lines (for winches)

On most high rises, the contractor will have concrete anchor straps about 4' long; they have a cloth loop on one end and a metal ring on the other. Ask the safety man for four.

1. Connect two straps together by the cloth loops, ending up with two straps — 8 foot long with a metal ring on each end.

2. Slip one end of each strap (metal ring) onto the hook from the winch.

3. Attach a large spring link hook through the cloth loop in the center of each strap.

4. Slide the metal ring end under the railing cap and clip it to the large spring link hook. (repeat with other strap)

5. Attach smaller spring link hooks above the larger hooks to connect tag lines.

The smaller hook will avoid confusion when connecting the load.

6. By adding two lower straps, you can lift an additional piece.

J. Making Your Own 5/8" Rope Spreaders

Use approx. 30' of rope. Attach a large spring link hook to each end. Add two more hooks 18" to 2' up from the first hooks, spaced as equally as possible. Add a final loop & clip in the exact center of the rope. Clip the center hook to the winch shackle. Pass the hooks on the other end under the cap near each end of the rail and clip it to the second hook without touching the rail as it rides. The hooks will quick connect and disconnect.

The tag lines are next. "Do not" use rope less than one half inch thick, it is too hard to handle. Polypropylene rope 1/2" x 600' is lightweight and inexpensive.

Attach one tag line to each end of the rope spreaders, just above the second hook. The tag lines will stay with the spreaders and the men on the ground will have control up, as well as down.

Chapter 4
Field Layouts

A. Core Drilling Layouts

1. Start by marking the slab, at center of each post with a one—two inch line.

2. Check the plan for centerline of railing set back, place set back marks at each end of balcony.

3. Determine the hole size to be drilled, 1 1/2" post needs a 3" hole, 2" post needs a 3 1/2" hole, 2 1/2" post needs a 4 1/2" hole.

4. Divide the hole size in half, a 3 1/2" hole divided in half is 1 3/4". Measure out 1 3/4" from set—back line, mark and pop a chalk line across the balcony. This will mark the front of the holes — keeping the holes in a straight line.

5. Next, measure over 1 3/4" from each side of the centerline of the post mark. Draw a line 3" or 4" long from the chalk line back. This will form an upside down "u" marking three sides of the hole.

Tips to help

Always use permanent marker to mark the centerline of the post. Have another color marker with you to correct any mistakes if necessary. Use a carpenter pencil, cut a square block the size of the core bit, and mark the center of the block. Match the centerlines and mark both sides of the hole. Set up the core drill with the bit in the "u" so the bit hits the edge of all three lines. The holes will come out perfectly aligned every time.

Laying out long runs with a measuring tape

Fiberglass tapes stretch and their accuracy is compromised.

Always use steel—measuring tapes. If there are many typical balconies, mark the post spacing directly on the 100' tape with a permanent marker. Then mark the first post, place the end of the tape on the mark and use a weight or concrete block to hold the end of tape in place, stretch out the tape and layout the rest of the run. If the run is longer than 100' find the center and measure out both ways.

B. Laying Out With Templates

Templates can be built from most any material found on a job site. Plastic corner—bead, electrical conduit and metal studs work well because they're 10' long. Wood is easy to work with and aluminum can be used for custom templates.

First, cut the material a few inches bigger than the railing. Then mark the centerline of the post with a permanent marker or two pieces of tape with a 1/8" gap between at the centerline. Make sure the center of the template is marked if there is not a center hole on the balcony.

Find the center of the balcony by sliding the template to the wall or slab edge and place a small mark on the slab at the center of the template line. Then slide the template to the other wall or slab edge and make a small mark. This will leave two marks 6 to 8" apart, place a bold mark between these two marks at the center of the balcony. This can be done by eye. Match the centerline of the template with the centerline of the balcony and mark the slab at the centerline of the post with a permanent marker.

Tips

Electrical conduit is easy to carry, can be used in tight spots when safety rail or other obstructions are in the way. It can be bent to handle simple radius work without much flexibility, and being ten foot long, it allows the layout of balconies up to twenty foot long.

Plastic corner bead can also be used. It is easy to cut, flexible to fit in the elevator, it doesn't roll and isn't dangerous if it falls over the side.

Half size templates to mark centerline of post

Cut the material to a little more than half the size of the piece. Layout the centerline of the post, with an equal space on each side. From the center of the balcony slide the template to one edge of the balcony, place a small mark. Slide the template to the other edge of the balcony and make another small mark. Between these two marks is the exact center of the balcony. Mark the center post, put the template on the center mark and layout one side of the balcony. Then slide the template to the other side and repeat the process.

If the balcony does not have a center post, layout the template from the center back, so one end has a half post space. Find the center of the balcony the same way, layout one side, and then flip the template end—over—end and layout the other side. I have included a diagram to help explain templates.

Plywood templates

Thin plywood templates work great on clean balconies or pare-pit walls. Layout the post, and then cut squares the sizes of the core bit. Rip the plywood to the size of the wall, or with the edge of slab setback in mind, so all four sides of the hole can be marked. Layout, is the same as discussed previously.

Templates with post

Template with post consists of a top cap with welded post, 6" to 8" long. Easy to handle, no finish to worry about, and can be passed from floor to floor.

1. Place template in desired location

2. A 2" post has 1" to the center of post. By placing a piece of 3/4" material against the post and drawing a line, this will mark the edge of hole location for a 3 1/2" core bit. Repeat on three sides to form a "u" to drill in.

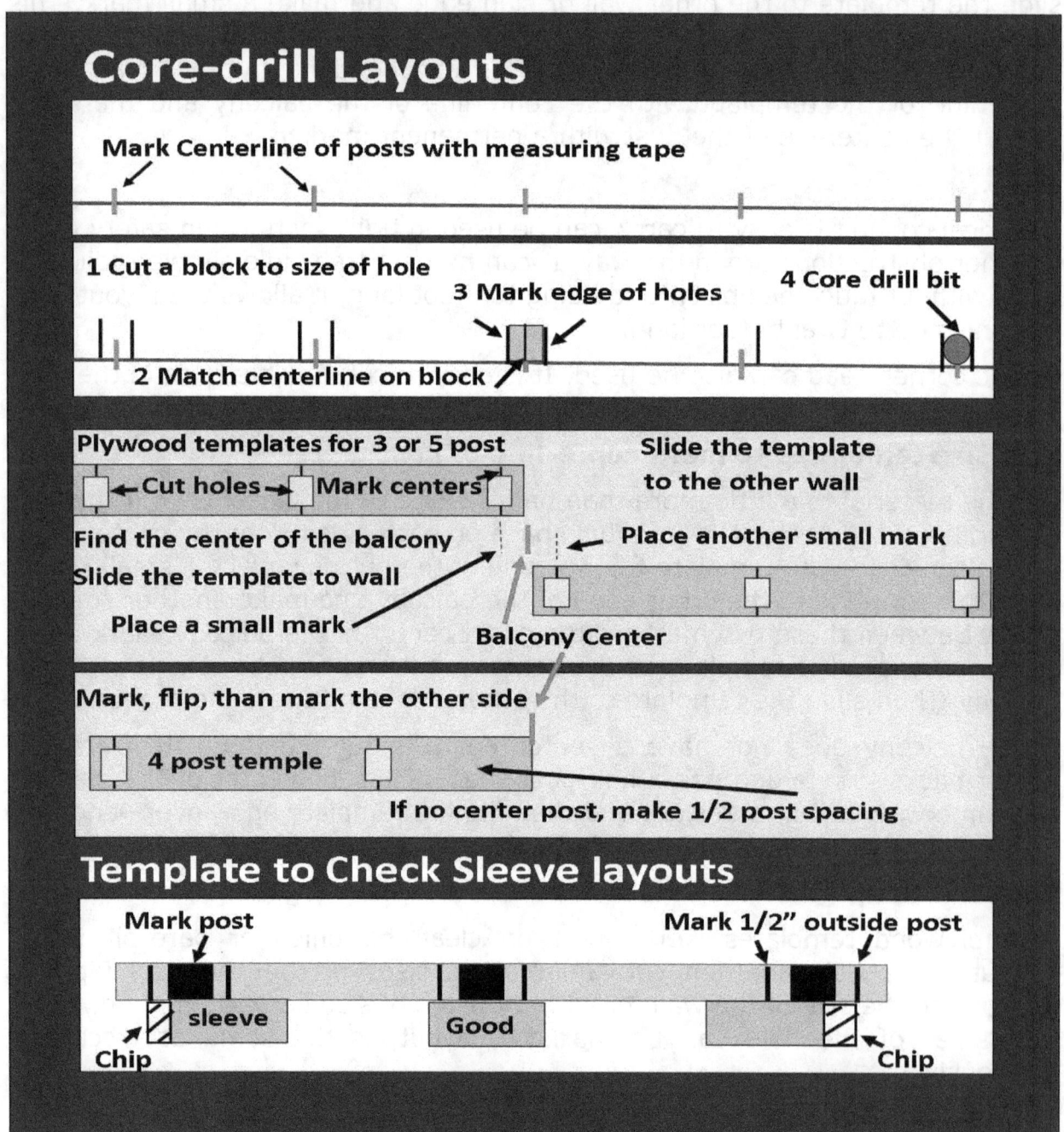

Templates with 45 degree post

There are times when there will be one or two, 2" 45 degree post on a balcony with standard 2" post. The 45 degree post will fit tight in a 3 1/2 " hole; it must be laid out exactly.

Cut an old 3 1/2 " core bit in 5" long sections, wrap tape around the 45 degree template post, and force the core bit sections over the tape onto the post. The tape will keep it centered. Add a screw for safety. Lay it out with a circle.

C. Layouts For Chipping Sleeve Jobs

Cut template a few inches longer then the piece. Layout the posts completely by marking a black square or with tape. Draw a line 1/2" out on each side of the post mark. Draw another line 1" out. Lay the template in place and mark the holes that have to be chipped. Use the 1/2" line if you are chipping. If the contractor is chipping use the 1" line, it will give you less to chip later. This formula will work with all templates.

Laying out with a measuring tape

If there are several typical balconies, blacken your tape at the location of the posts with a permanent marker. It sometimes pays to use cheap measuring tapes for high—rise work.

Tips

If possible, check the sleeves after the first floor is poured. Make all necessary corrections clear to the contractor. Do whatever it takes to help the contractor set the sleeves accurately. Make special layout templates, special measuring tapes, discuss the process with the person actually setting the sleeves. If this part is done right, the job will go smoothly and everyone will make money. If the sleeves are set wrong, even if the contractor is responsible, everyone will lose money.

D. Laying Out Fences In The Dirt

Run perimeter string lines to mark the centerline of the posts. Measure out the post spacing and mark the centerline of the posts both ways with a cross on the ground with spray paint, to be dug with a posthole digger.

Chapter 5
Core drilling

A. Core Drilling

Always core drill with a helper whenever possible. He can layout ahead, extract cores, and chisel out cores that do not break clean, wash down, help pop chalk lines, reroute water and power, and remove obstructions ahead.

Typical scenario

Know in advance, where the water and electric are located.

Bring enough hose, a hose "y", extension cords, a three—way plug, and a water hose pressure reducer, from an R.V. store.

The focus is on keeping the core drill running. When the core drill is not running, time is lost that cannot be made up.

Bring equipment to balcony, set up water and power.

With your helper, pop the chalk line and layout the first run. If the run is too long for a template, mark the centerline of the posts with a tape. Complete the layout of the first hole, set up the core drill, always brace to the ceiling with a 2x4, and tie the 2x4 off to the rig with a thin rope, for safety. Start drilling. Once the core drill is running, the helper is free to layout the rest of the holes.

After the first hole is drilled, quick disconnect and drop the hose. Setup on the next hole, wash down the previous hole and reconnect the hose. If you are not in direct sun light, wash down three or four holes at a time. Helper can drop back and extract the cores.

Before the man core drilling is finished with the run, the helper can start laying out the next run. When the time comes to move to the next run, the helper stops whatever he is doing, and helps move equipment. Whatever time he loses can be made up. Sometimes the helper has enough time to wash down.

B. Core Drill Rigs

The best core drill rigs are Hilti. They cost about a thousand dollars more, and are well worth $10,000 more.

Order the most heavy duty one they have.

Order a narrow base, so it fits easily on stairs and in corners.

Order a screw jack for an overhead ceiling brace.

Order a threaded adaptor to accept standard core bits.

Order an electrical adaptor from twist grip to standard; in some states twist grip is standard.

Everything is setup to quick disconnect

The motor quick disconnects from the stand and is easily to carried in two pieces.

The water is also set up for quick disconnects.

The handle has four spokes and easily connects to either side.

The handle can crank at a one—to—one ratio for concrete, and quickly adjust to three—to—one ratio to cut through steel.

The screw jack has a half wheel, with a swivel tip to the 2x4 ceiling brace.

It has an adjustable depth gage.

Quick disconnect core bits are also available.

Water catching corrals are available. They improve every year.

Hilti core rigs are very cost effective, the setup and drilling time are minimal.

If you already have a core drill, it can be modified.

Your stand should already have a jack bolt on top. Have a half wheel welded to it, so a wrench is not necessary to turn the bolt to brace to the ceiling.

If the core drill has a slide bar handle, center the bar and tighten a zip tie on each side of the bar to keep it from sliding too freely. You will still be able to force the slide bar when necessary. Be creative.

Make a depth gage or stop, out of wood, electrical conduit or other material. With the bit resting on the slab, measure the distance between the base and the yoke; minus the desired depth of the hole, and cut the stop accordingly. Tie wire the stop to the column, so when the desired depth is reached the yoke will hit the stop. Tie wire to the stop in the center, so the stop can be twisted out of the way, to finish cutting through a piece of rebar, if necessary.

Always drill 1/2" deeper than the post that will be imbedded, to aid in adjusting height.

Find quick disconnect hose ends with a shut off. They are sometimes hard to find. Ace hardware will have them, Home Depot or Lowes will not.

C. Core Bits And Core Bit Modifications

Core bit teeth are made of carbide mixed with diamonds.

Core bits are priced according to diamond density in the teeth.

The type of concrete mix determines the necessary density.

Concrete mix is determined by what is plentiful in that area.

For example south Florida concrete mix, is shell rock and sand. Whereas in Georgia, it's granite and sand. The hole that takes 90 seconds to drill in south Florida could take 25 minutes in Georgia. The harder the concrete, the faster the bit wears out.

Diamond Products color code their bits, black for the highest diamond density, white for the lowest.

For real hard concrete, they make teeth with high diamond density in a softer carbide tooth. They are expensive, they cut through anything, but they do not last long. (40—50 holes)

Short 9" bits can be ordered through your supplier. They lower the center of gravity on your machine and make it easier to handle, move and setup.

Core bit modification

When the core bit is cutting through rebar, sometimes it will hit what is known as edge steel, steel that runs alongside of the hole. The core bit cuts the edge off the rebar. A little sliver of the rebar stays in the core. Then it falls between the core and the bit, jamming the bit in the hole. This can be a major problem. By working the bit back and forth with a wrench on the bit, it can take as long as 45 minutes to free it. If the bit is less than 3 1/2", just tighten the crank, pull back sharply on the rig, the core should break off in the bit, and can be chiseled out. This causes undue wear and tear on the rig, the larger the bit the more difficult it is to get it free.

The solution is to cut one tooth off the bits. Mark the side of the bit clearly, so you will know exactly where the missing tooth is when the bit is in the slab. When it jams look closely at where the bit enters the slab, the bit should be pressed to one side of the hole. Turn the bit with a wrench until the missing tooth is at the tight side of the hole. Slowly turn and gently crank the bit out of the hole. If it is not clear where the steel is, the rebar usually runs along the slab edge on the inside or outside of the hole. Crank the handle gently and turn the bit a little at a time. Once the bit is free, crank the motor to the top and extract a core. Remove the sliver of steel with a magnet and finish drilling the hole. Most of the time the ceiling brace will not have to be removed.

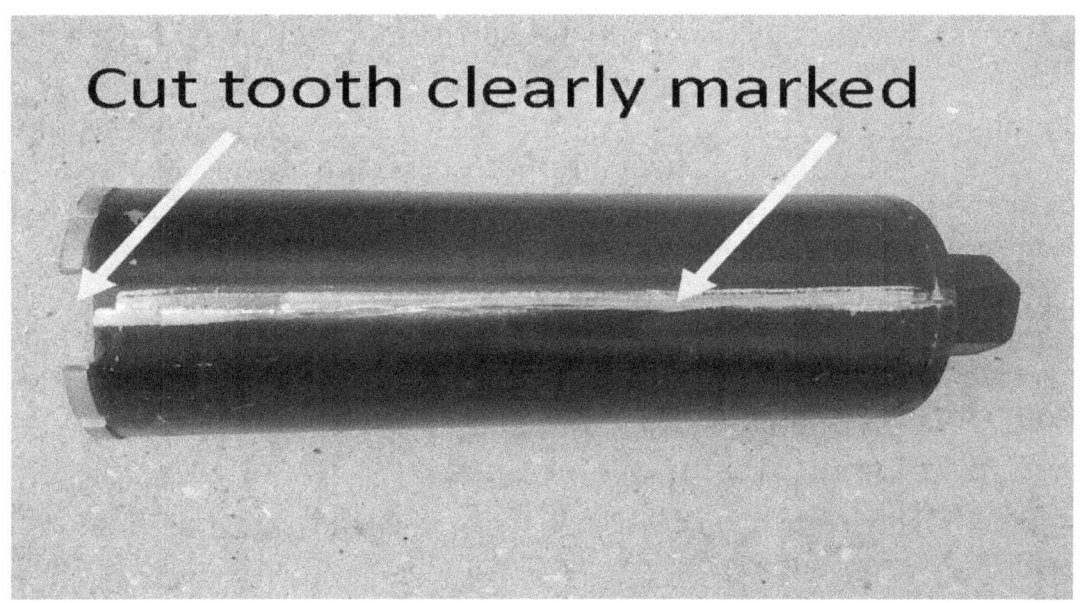

Cut tooth clearly marked

D. Core Extraction

Extract cores by placing a chisel in the slot and striking it sharply with a hammer. Never place the chisel between the core and the slab edge, as slab edge may crack, or even blow out. This is especially important when drilling parapet walls. The deeper the hole, the easier it is to extract the core.

If the hole is not deep enough to extract the core in one piece, after freeing a jammed core bit, chisel the portion of the core where the steel is and remove the sliver of steel with a magnet. Finish drilling the hole, the core will extract normally.

When extracting large cores, take a small chip out of the core, place the chisel in that space, then extract the core, it will pop without damaging the slab around the hole.

If the core pops but will not come out it, may be that a piece of up steel or tie wire is embedded in the core. Split the core in the center and take it out in pieces. If the core does not break clean, then chisel out the concrete by starting in the center and working outward.

If there is rebar coming, up from the bottom of the hole that the core bit does not cut or the rebar hooks, it can be cut with a torch. If a torch is not convenient, and the rebar is 5/8" or less, it can be cut with a high—speed Dremel. Chip away the concrete from the cutting area, cut into the rebar approx. 1/4 of the way, with a reinforced cutting wheel. Then bend the rebar back and forth by sliding a piece of pipe or electrical conduit about 12" long over it, for leverage. It will snap at the cut. Paint or epoxy the steel to keep from rusting.

The aluminum post should never touch steel inside the hole, as the dissimilar metals with moisture will cause internal slab damage. A segment of the bottom of the post, near the steel, may have to be cut away to keep clearance between the aluminum and the steel.

Many times, we have the opportunity to drill parapet walls before they stucco. Keeping stucco out of extracted core holes can present a problem. Clean sand is the answer. Fill the empty holes and let them stucco right over it. The holes will clean out quickly and easily.

E. Flow Control And Catching Water Slurry

Flow control starts with a quick disconnect hose end that has a shutoff. Not just for flow control, but for easy wash down and more efficient moving of the core drill.

Most core drills already have some type of flow control. Set the flow control on the core drill to desired flow and set it there; the quick disconnect hose end can turn the water on and off. If the building is not painted and running water is not an issue, run the water fast enough so the slurry does not paste up and is easily washed down.

Never core drill over finished railing, or anyone else's product, whether installed or sitting on the ground below, unless you are catching the water. The water slurry carries steel filings when cutting rebar. The steel filings will rust wherever they land.

If drilling in an already painted area and you are not going to catch the water, wet the painted surface thoroughly before drilling. Slurry is less likely to discolor a wet surface. Run the water fast, so the slurry dilutes as much as possible. Wash down quickly

Water diversion can be very useful to control the direction of the slurry. This is accomplished by rolling up a rag, and wrapping it around the bit, but not touching the bit, letting the tip of the rag hang over and a little away from the slab edge. The slurry water will run off the tip of the rag.

Unable to get a water hose to the drill: rig a 2 or 3 gal. Insect sprayer with a hose end, and a quick disconnect. It will drill three or four holes.

If you are catching the slurry, run the water slow, but fast enough not to clog or bind the bit.

Several ways to catch slurry; Use a Hilti water collector corral.

Wrap a wet towel around the bit, leaving a 1" gap between the bit and the towel. This will create a little puddle around the bit. Turn the water off before it overflows. When the slurry starts to get pasty, sponge it out and add more water.

Catching water with a wet vacuum; Have a helper hold a wet vac crack attachment, against the edge of the bit and the slab, sucking the water as it flows. The bit will wear a round spot in the tip of the crack attachment. This will help the process. If a little water gets away from you, let it go. If you try to get it, you will lose more.

Making your own water corral: purchase a 4 1/2" x 6" flush bushing from a plumbing supply house. Drill a hole in the outer housing so the bottom of the hole is about 3/4" above the slab. Insert a 1 3/4" threaded nipple, use two thin electrical conduit nuts, one on each side of the outer housing, if necessary. Tape vacuum hose or the inside of the nipple, building it up, until the vacuum hose can slip on with a tight seal.

Grind the webs that hold the inner and outer housing together, up about a 1/2" so they don't set directly on the slab. Then fill the chambers, except the one with the threaded nipple, with anchoring cement, leave 1/2" clearance between the anchoring cement and the slab. This will add weight and stability.

Tips: Kmart carries a wet vac that pumps water out through a garden hose.

The only down side to wet vacs are, the core drill and the vac have to run at the same time. If the circuit breaker blows, the vac and drill stop, but the water keeps running.

The circuit breaker on a generator sometimes blows.

However, power drawn from a swing stage should be sufficient, or running two cords, from two different circuits will work.

F. Core Drilling Parapet Walls

Parapet wall heights and widths vary greatly; from a 6"x 6" curb to a wall 36" high x 12" wide.

If the wall is low and narrows enough, the core drill can set on the slab, and a short bit can be used.

Two platforms, the height of the wall can be built; one to use and one to move to.

If the wall is block with a poured cap, the narrow base core drill can sit on the wall sideways and brace to the ceiling. Tie off the rig from above for safety, preferably the roof. The higher the tie off line the greater the swing, the more holes can be drilled without moving the rope. (More than one rope is always an option)

I created a wall protector—easy slider. Start with a 3/4" plywood top, with 2x4 sides, so it can slip over the wall for the core drill to sit on. Add four hard plastic easy glide floor protectors to the bottom of the plywood, one in each corner to add stability, to an imperfect wall. Carry a door shim and slide it under one corner when necessary to add more stability.

After drilling the hole, hold onto the core drill, and slide the wall protector and core drill backward, to drill the next hole.

Important: The core drill turns clockwise, when it binds it will try to swing the base off the wall. Make sure the base swings in toward you and not out away from you. You will have to make an exception on the last hole.

Wall protectors, can also be used on narrow walls. Be sure to tie off the core drill, because the plywood and core drill will overhang the wall on both sides.

G. Core Drilling Without An Overhead Brace

There are times it will not be possible to brace overhead, such as, around pools, site rail, or stair treads.

Set up the core drill, and get the hole started when the bit is in the concrete approximately 1/4", have a helper sit or stand on the drill motor for weight. Use the crank to control and relieve the down pressure when necessary.

Rig a 50 lbs. Counter weight or a bucket of anchoring cement on the motor. Use caution when moving. A short bit will add stability.

H. Core Drilling Through Marble Or Tile

Getting a hole started, and maintaining accuracy on marble or tile is difficult, the bit has a tendency to wobble on smooth surfaces and damage the floor.

The bit can even wobble out of control if not braced overhead.

If your core drill, drills true, with no wobble; layout the holes on top of masking tape stuck to the floor. Slowly drill through the tape until the bit starts. The tape will help stabilize the bit until it starts.

If the core drill has a wobble to it, (most do) or you do not want to take any chances; setup the core drill on a thin piece of plywood. Dry drill a hole through the plywood, cut the plywood in a strip, with the hole on one end, so you can place the hole in the wood over the desired location on the marble. Have a helper stand on the plywood with both feet, so it does not move, while the hole is started.

I. Hand Held Core Drilling Rigs

Hilti and other companies are making hand held rigs. They are lightweight, easy to handle.

The down side, there is no crank or overhead brace; it is all manual push, and not recommended for drilling too many holes.

Your supplier can order an adapter that fits a standard core bit on one end, and 5/8" threads that fit a 7" grinder on the other end. The RPM's have been geared down to a low speed and it has a water connection in the middle. It works well in all applications.

Diamond Products make a dry core bit with 5/8" threads, that also fits a 7" grinder. It does not cut steel, so its application is limited. It works well when removing old railing. Cut the post, leaving an inch or so coming out of the slab. Use the smallest bit you can that fits over the old post. The cut post will hold the core bit in place while drilling.

To get a hole started without the help of a cut off post; cut a square hole the size of the bit in the end of a strip of plywood. Place the cutout over desired location and stand on the plywood to keep it from moving while starting the hole.

Chapter 6
Railing Installation

A. Balcony Railing Codes

Minimum— 42" high; a 2" maximum space underneath, a 4" maximum picket spacing.

Inspectors interpret this as being 42" high from the slab to the top of the railing, and 4" from the wall to the post, so a 4" sphere will not fit through.

From the bottom channel of railing to the slab, is 2" maximum; this is to keep anything bigger than 2" from rolling off the slab. If the railing is set on a curb, the 4" rule applies.

Railing around pools varies from 4' to 5' high. Picket space is a 4" maximum. From bottom channel to ground is also a 4" maximum.

The gate must have spring hinges with an automatic locking latch. It must open out, so if the gate does not latch, a child approaching the gate will push the gate closed rather than open. In some cities, the code is that the latch must be 5' high. They now make a latch that extends above the gate, which can be added to existing commercial pool gates to bring them into code.

B. Connecting

Railing is typically designed to be connected at a post. The first piece will have a post on both ends and all the other pieces in the run will have a post at one end and be open at the other end.

The top cap of the railing should be designed to splice in the center of the post whenever possible. This helps the two caps to join properly, especially if the railing is made at a radius, or the inter splice does not fit tight enough. If the inter splice fits tightly, the top cap can splice at the edge or off the post between the post and the first picket. The best—case scenario is the top cap splices in the center of the post with a tightly fitting inter splice welded into the cap at the post end of every piece, the inter splice is painted with the railing. If there's a small gap in the joint, it would be acceptable as an expansion joint. If the inter splice is added in the field, install the inter splice into the post end of each piece with a stainless steel tech screw through the underside of the cap, between the post and the first picket. The center of the inter splice must be painted before installation. If the inter splice is not painted and silver shows through the joint in the cap, it will be considered a mistake instead of an expansion joint, and will have to be caulked.

If the two caps do not join properly, and there is a gap on the top, bottom or side, the open—ended cap will have to be cut or ground back until the caps match. This happens more on radius railing and corners, then straight runs. Measure the open gap, then measure back the same amount at the point that is touching. Place a mark with an ultra—fine point Sharpe pen. Turn your measuring tape over (numbers down—creating a flexible straight edge) stretch it from the mark across the top cap to the edge of the cap at the widest point, draw a line, then cut or grind to the line and the caps will fit properly.

Start with the piece that ends with a post, install the inter splice, if necessary. Place the piece in the holes with bottom channel on top of 1 1/2" spacers, (2 x 4 blocks) one at each end of the piece. This lifts the bottom of the posts off the bottom of the holes, so the railing can slide easily, taking advantage of any side—to—side play in the holes. The spacers will be used later when leveling.

Bring the next piece with an open end into position, next to the first piece. Lift the post end of the first piece with the inter splice attached. While your helper is lifting the post end of the second piece, slide the open—end top cap onto the inter splice of the first piece. The 1 1/2" spacer on the other end of the first piece will enable the first piece to be lifted high enough, so the bottom of the post on the second piece will clear the slab and drop into the holes.

The bottom channel of the second piece will slip over a clip pre—welded to the post and butt against the post. Tighten a 1" ratchet strap from post to post just above the bottom channel, to keep the bottom channel against the post while connecting the next piece. Continue process until the run is complete.

Do not screw joints together until all the pieces are in place, and within leveling tolerances.

When a piece is made with both ends open, you will have to connect the post end to an installed open—ended piece. The open—ended piece will have to be lifted much higher to allow the bottom of the post (on the post ended piece) to clear the slab and drop in the holes. In order to lift the piece high enough, a bigger or double spacer will be required at the other end of the piece that is already connected. The piece it is connected to might have to be lifted a little higher, as well. Ratchet straps are critical in this situation.

Sometimes it is necessary to set a partial run.

Never grout the last piece. Be sure it can be lifted high enough to set the next piece, before grouting the pieces before it.

Tips and tools

If the inter splices are not welded in, fit the railing properly before installing the inter splices. Aluminum can be easily cut with a circular saw using any plywood blade sprayed with WD40; most other spray—oils cause the blade to clog with aluminum. Blades with 60 carbide teeth last much longer, because they cut clean and do not fill up with aluminum.

Non—carbide metal cutting blades are too thick, do not cut clean, and fill up with aluminum very quickly

When grinding, use only aluminum wheels, not steel. They cost more because they are worth it.

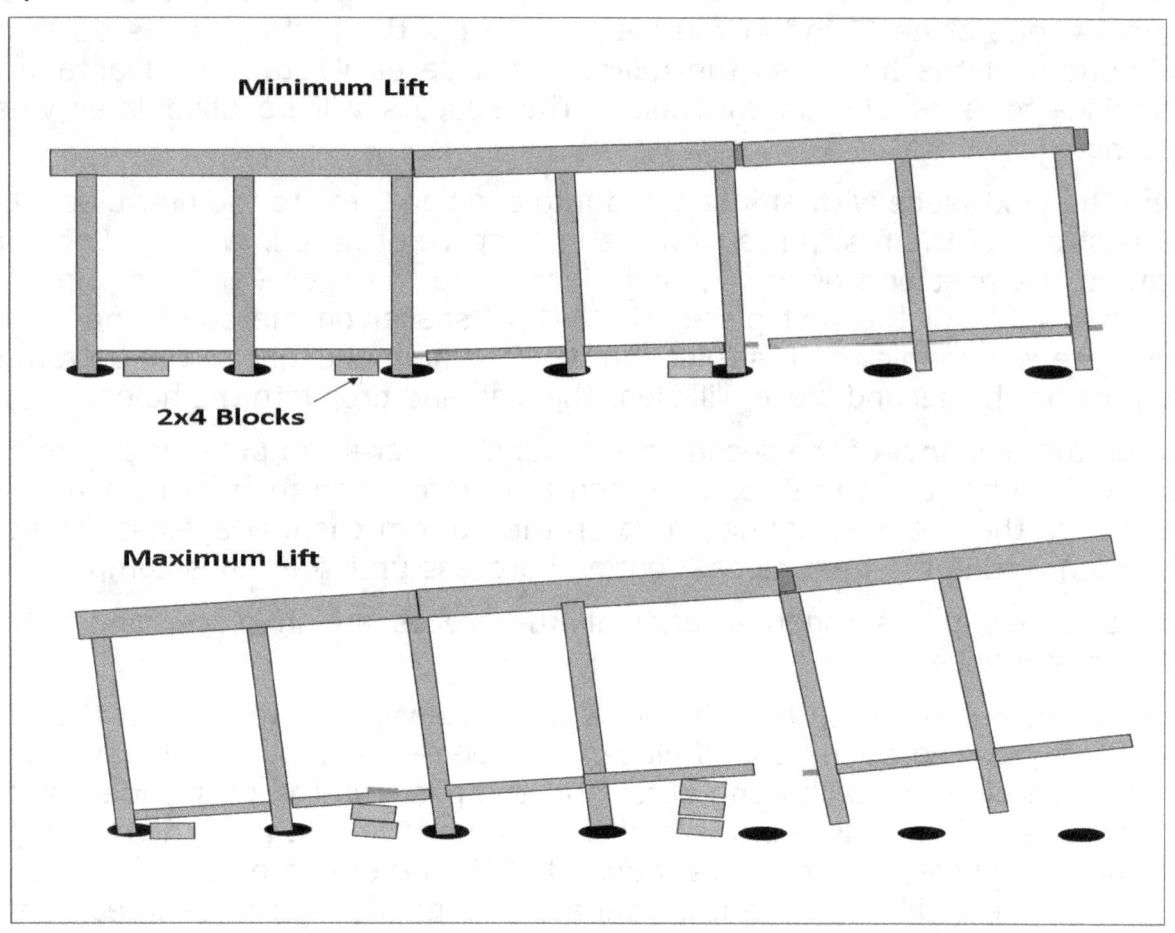

C. Leveling

The railing should be manufactured to allow a 1 1/2" clearance between the bottom channel and the slab, and a minimum of 43" high.

This enables the railing to be leveled, while maintaining height and slab clearance codes. Concrete slabs are imperfect and can have severe high points and dips. Balconies with returns can be pitched as much as 2", which will have to be addressed when the railing is manufactured.

Start by placing 1 1/2" spacers between the bottom channel of the railing and the slab on the post side of every joint, near the post but not covering the hole. If the railing is lightweight or mechanical, a spacer may be needed at every post. Cut 2x4 blocks, approx. 4" long, to use as 1 1/2" spacers. If you split the blocks, you will have twice as many. Different size spacers, 1/16", 1/8", 1/4", 1/2", 1", 1 1/4" and door shims will be helpful.

Eye the top cap of the rail (long ways) and start at the highest point.

Replace the 1 1/2" spacer with a 1" spacer and use whatever other spacers are necessary to level the rail.

Place a torpedo level on the picket side of the post, to determine which end of the rail to lift or drop.

If the balcony is facing the ocean, eye the top cap against the horizon to determine which end is high or low. Add and remove spacers until the railing is level.

If a high spot persists after the 1 1/2" spacer is removed, it's because the post is hitting the bottom of the hole. The post might have to be trimmed, or the rest of the railing might have to be raised. Keep in mind, no more than 2" under and no less than 42" high.

If leveling and meeting code requirements are going to be a problem, keep the high at 42" and work both ways, making a gradual transition to the high spot.

Before doing this, let the contractor know his options are: grind down the high spots on the slab, add concrete to the to the low spots, disregard the 2" rule and add flat bar under the bottom channel or live with high spots and dips in the rail.

D. Aliening Or Plumbing The Railing

Now that the railing is level, cut 1x2 or 2x2 wedges approx. 6" long, depending on the size and shape of the posts and how much clearance between the post and the edge of the hole. Cut 2x4 wedges approx. 12" long, for free—standing pieces. Next, cut props about 43" long, depending on the size of the cap. Collect pieces of coated solid core copper wire to be used as tie wire. Solid core, so it can be twisted. (strand wire has to be tied) Coated, so it does not scratch the railing and can be reused.

Single Freestanding Railing — 20' or less, That Sits Between Walls:
With one person on each end, center the rail between the walls; place 2x4 wedges between the wall and the rail on each end. With a torpedo level against the inside of the post, plumb the rail; strike the wedge with a hammer — this will tighten both ends. If both ends will not plumb, average out the error and strike the wedges both ends for added tightness.

Two or More Free Standing Pieces of Railing Between Two Walls:
Plumb both ends using 2x4 wedges. Eye the railing — if the railing is pushed out at the joint, use a 2x2 wedge in the hole to align the bottom of the post. Use the tie wire or ratchet strap to draw the railing in. Push the wire through the drain slot on the sliding glass door track — twist it in place. Rap the other end around the top of the post just under the cap, pull the top of the railing until it leans in slightly, and twist the wire to hold the rail in place. Now, slide the wire down the post until the joint is aligned to perfection.

If there isn't a sliding glass door track in the necessary position, hook the top corner of an electrical box with one end of a ratchet strap and hook the other end around the top cap and draw it in. If there is nowhere to hook or twist, use a full bucket or two of anchoring cement to hook to.

If the railing leans in, place a prop against the post (under the cap and to the slab) to push the rail out. The prop should hit the slab about 2' from the rail, to avoid pushing up — instead of out.

How to plumb freestanding railing in windy conditions: Stabilize the end post; pull the posts in with wire, while pushing them out with a prop. Once they are plumb, it might be necessary to set them with anchoring cement before proceeding to plumb the remainder of the post.

If wire or props are not appropriate, place a 2x2 or 2x4 wedge in the hole on the inside and outside or the post. Move post into plumb position and strike the wedges with a hammer. Once the wedges are tight to the post, the post can be fine—tuned by applying a little inward pressure and striking the outside wedge or outward pressure ad striking the inside wedge. The post will only stay in place if there is not too much pushing or pulling elsewhere. It may be necessary to set the post with anchoring cement before continuing. This is a good method for short single pieces.

Free standing runs— more than 100' to 200' long: Plumb and set both end posts with anchoring cement. Set up a laser torpedo level at one end of the railing. Build a stand so the laser can sit in the center of the rail, without touching the cap, if possible. Clamp a small reflector target to the center of the cap on the other end of the run. Adjust the beam to hit the target; this will require binoculars. The beam will be about an inch above the top cap. Place a centerline on a small piece of wood the size of the cap, starting with the joint that is furthest out of alignment; bring it into alignment, use the small piece of wood on top of the cap as an intermediate target to check alignment. Repeat the process on the next joint that is most out of alignment; repeat until rail is perfectly aligned. If by moving one section into alignment, another section is pushed out of alignment, set the posts that are the most out of alignment, after anchoring cement is hard and then continue setting the rest.

Remember it is easier to push out with a prop, than pull in with a wire. Set the post that had to be pulled in with wire first, and then push out the post that is leaning in with props.

Setting single posts: There will be times you must set a line of single posts for a variety of reasons. Such as, wood cap to be set by others, glass panels that set between the posts, or the cap may be a different color than the post and mechanically fastened later. The post must be set plumb, all the same height without variation and in perfect alignment. If there is any adjustment, it will be very little.

For alignment, pop a chalk—line if possible or invest in a torpedo laser level. For height, a rotary laser is best; a laser level can be used after setting the first and last post. Sit the laser on the first and the target on the last. If using a string attached to the first and last post, use a spacer to keep the string above the railing and measure down. Sometimes the string can be attached wall—to—wall above the railing. Use shims and spacers under the posts to bring them to the proper height, wedge each post into plumb and alignment.

Important: Pouring anchoring cement is a two—phase process. Do not try to do it in one. In tight holes, using small wedges, keep the anchoring cement down an inch or two from the top of the holes, on the first pour. This will allow for easy removal of the wedges vs. digging out broken wedges with a hammer and screwdriver that cannot be used again. (see pouring AC)

Chapter 7
Stair Railing

A. Stair Codes

1. Stair railing codes are 42" high

2. No more than 4" gap or picket spacing

3. No more than 4' between posts

4. 44" minimum travel space from post to wall

5. A 6" ball cannot roll off the side of the step. (bottom channel clearance 1 1/2" or less should prevent this)

B. Typical Stair Layout

Whenever possible, layout stairs with the actual railing. Place all the pieces in their proper locations, starting at the top landing. Make the pieces plumb before marking where to drill. Measure 4" back from the stair nose, this marks the top post of the first piece. Place your torpedo level on the side of the top post to see whether the piece is leaning forward or backward. Lift the low end until the piece is plumb, a wedge under the post will hold the piece plumb, while you measure the space between the bottom of the lifted post and the ground. If the space is less than 1", trim the post touching the ground and the center posts, as well. Mark the center of each post, continuing all the way to the bottom. The distance between the side—by—side landing posts should be predetermined by the gap between the stairs. Make a wooden template to mark the landings, as the holes will remain standard.

Rack the piece when possible. If the top post is leaning forward, pick up the piece and strike the bottom of the top post against the nose of the top step. If the top post is leaning back, pick up the piece with the toe of the bottom post on the landing, lift and drop, and then check for plumb. Caution; do not damage the railing. This might cause the cap to bow down slightly in the center. Plumb and set the top and bottom posts with anchoring cement. After the anchoring cement hardens, wedge up the center post, hold it in place with a spacer between the nose of the step and the bottom channel, and set the center post.

If it is necessary to core drill the stairs before the railing is fabricated, a layout formula will be necessary. Start with the return to the wall on the top landing. Assuming the post are going to be 2", place a mark 3" from the wall and 4" from the slab edge, that marks the centerline of the first post. Make the next mark 3" minimum from the side of the stair tread, in line with the first hole;

this will insure the proper setback of the top hole of the first stair section. Proceed down the stairs to the middle of the center stair tread and mark; keep 3" from the side of the stair tread to the center of hole all the way down. The bottom hole of the first piece should line up with the top hole of the second piece. The centerline of the top hole of the second piece should be marked on the landing 4" back and 3" from the edge of stair. Lineup the bottom hole of the first piece, mark 3" from the side of the stair. Continue this formula down the stair tower.

It might be necessary if the landing is short, to layout the bottom post of the first piece on the center of the last tread before the landing and the top post of the second piece on the center of the first tread after the landing. If possible, keep the holes in line if a filler piece is necessary. (The holes can be adjusted back away from the nose if needed.)

The project manager will be able to make the railing fit the holes.

If the stair is typical, a wooden template can be fabricated. After laying out the first piece, place a 1x4 or 1x6 the length of the railing, down the nose of the stair. Fasten pieces of furring strip to the 1x4, making sure they hit the top of the steps and landing, and are plumb. If the left and right stair is different, make two templates. Sometimes the second template can be built on the flip side of the first.

C. Core Drilling Stairs

Start at the bottom and work up, out of your own mess.

Run the water hose from the bottom up. The power cords from the top down to keep them from tangling.

Brace to the ceiling when drilling landings, and have your helper stand on the motor when drilling stair treads. They drill very quickly; there is never any steel.

To move from one hole up to the next, lean the rig sideways and lift the base while the helper is lifting the column.

Rinse as you go, and wash down thoroughly from the top down when the drilling is complete.

D. Stair Rail Installation

If the stairs are not wide enough to meet code, stair rail has to be side mounted. Start at the top, and work down.

Starting with the return, place it on 1 1/2" spacers, add shims to bring it level. Keep in mind, 2" max bottom clearance and min 42" high.

Connect the first stair piece to the return at 42"high min, measured from the nose of the stair tread vertically to the top of the rail. The space from the nose of the stair tread to the bottom channel should not exceed 1 1/2".

Bring the bottom of the piece into plumb with spacers and shims, while keeping the return piece plumb. Once connected, the two pieces should stand on their own. If there is a filler piece, place it on a 1 1/2" spacer, bring the stair piece into plumb, and attach the filler piece. Attach the next stair piece to the filler piece — continuing the process all the way down. Recheck for plumb at the landings. Props to the wall may be necessary to keep the railing plumb.

If there is no gap between the stairs a filler piece will not be necessary. Cut wood blocks 3 7/8" long, place one on the landing between the two posts and hold another near the top of the post. Clamp the two posts together between the top and bottom block, with a 12" quick grip clamp. The blocks will maintain a less than 4" code separation.

Tips: setting the railing with anchoring cement should be done after working hours, to insure against outside interference while anchoring cement is drying.

E. Installing Grab Wall Rail

Codes: height varies city—to—city. Check the plan.

Typical height is 34" min — 38" max.

Diameter is less than 2".

1 1/2" from the wall.

It must break level 12" at the top and bottom landings.

Measuring for wall rail

To get the length without the 12" level breaks, measure from the nose of the top step down to the landing. Then take the angle.

If the rail were to be laid on the steps, without mounting brackets, it should lay flat 12" on the top landing. Continue down touching on the nose of every step, lying flat for 12" on the bottom landing.

Wall grab rail stands

Grab rail stands can be made of wood or metal. Start by forming a "T" as a base that will set on the landing, about 18" parallel to the wall, and 10" out. Fasten a piece of wood or metal inside of the "T" that comes up about 40" high. Determine the height of the grab rail.

Fasten the last piece of material to the upright at 1 1/2" below the height of the grab rail, coming out about 6" toward the wall. Make two stands, place one on the top landing and one on the lower landing. Set the grab rail flats on the stands. Then slide the rail until the angle breaking point is directly over the nose of the top step. Check the height by measuring from the nose of the second step straight up to the top of the railing. Check again from the nose of the last step. The height should be the same. The height from the nose can be raised by sliding the rail forward and lowered by sliding it backward. Use a shim or a spacer to raise one end.

Tips and tools on setting the wall grab rail on scissor stairs

Tape a torpedo level to a 3' straight edge and mark the desired height on the straight edge. Spread all grab rail with a piece lying on each stair. Start with a person on the top landing and a person on the lower landing. If the walls are block or concrete, use Tapcons 1/4" x 1 1/2" with a 5/16" hex head and 1/4" stainless steel flat washers.

With the stands in place, lay the rail on top. The man on top plumbs the straight edge off the nose of the second step, slides the rail until the top of the rail is on the height marked on the straight edge. The second man holds railing in place while the first man drills the upper most bracket. He then hands the drill to the second man. While the first man is screwing the first bracket to the wall, the second man is drilling the rest of the holes. After the first screw is in, the first man pulls out the top stand hands it to the second man and continues putting in the rest of the screws. When the second man is finished drilling, he leaves the drill at the top of the next landing and takes the stand to the lower landing. When the first man finishes screwing, he pulls out the lower stand and uses it at the top of the next landing. This process is continued all the way down the stair tower.

Tips on setting grab rail on long stairways

After the first man fastens the top bracket, he moves down to the center bracket. While the second man at the bottom is looking up the rail, the first man adjusts the center bracket. When the second man says it is lined up, the first man fastens the center bracket. Once the railing is in line, continue as previously described.

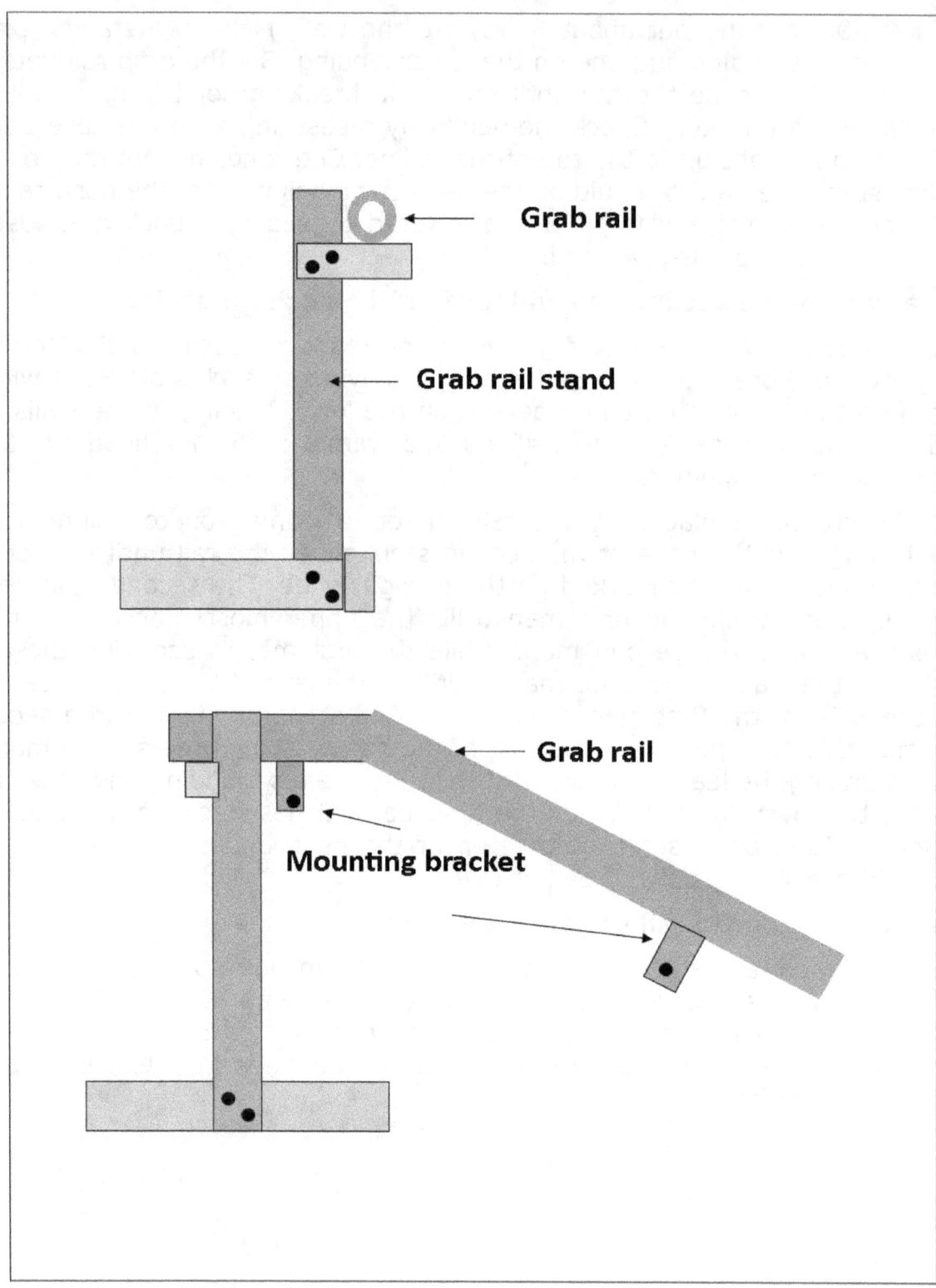

F. Continuous Grab Rail (every scenario)

Start by laying the grab rail on the stairs, connected together with an inter splice. If the brackets hit the stairs, set it up on the same size spacers — top and bottom.

The length, the break angle, the 90 degree turn must be accurate! Any error will accumulate. The taller the building, the worse the error will get. The space between the straight section after the 90 degree and the wall, should be 1 1/2" min and 2 1/4" max. Raising or lowering one end will open or close that gap. It is important to layout as many stories as possible to determine what adjustments to make and where. Record all adjustments, start at the top, and work down. Use your grab rail stands whenever possible, build more if necessary.

Continuous grab rail mounted to stair rail with loose transitions, has to be custom fit in the field

There are several variations of this setup.

1. The straight grab rail: the brackets and transition pieces are separate with two transition pieces per landing. Attach all the brackets to the existing stair rail at the proper height. Attach an inter splice to hold the two transition pieces together, and one extending from each end about 3". The two pieces should be able to swivel, be sure they send an extra straight piece that can be cut a little shorter. Start at the first landing down from the top, place the transitions on the brackets, slide a straight piece over the inter splices on each side. Move the transition forward or backward until the straight pieces lay in place on the rest of the brackets and the transition piece is in the right place to meet all code requirements. Attach transition pieces. The piece going up the stairs should break level at the top and meet height requirements. Pull it off the transition inter splice, lay it in place with the end overlapping and alongside of the transition piece. Mark it and cut it with a chop saw, slide it back over the inter splice and attach it to the brackets.

2. Now that the top piece and the first transition are set, move down to the second transition. Put all the inter splices in place, slide a straight piece on the transition going down, position the transition in its approximate location and measure the distance between the first and second transition, cut the extra straight piece an inch or two shorter. Slide it over the inter splices of the first and second transition, move the second transition exactly in place. Measure how long straight piece number two has to be and add 1/16" — cut the piece. Replace the shorter piece and with the piece cut to fit, put transition number two back in place and attach it. Replace the piece going down with the shorter

piece to connect transition number three and continue the process all the way down the stairs.

3. The next variation is one transition made to a straight piece that goes down the stairs and a smaller transition connects them together. You will need an extra small transition piece, which you can cut shorter. It is installed the same way as previously discussed. Just the transition piece is cut to fit. This system has fewer pieces and is stronger.

4. All other systems are similar. Be sure the manufactory sends extra pieces.

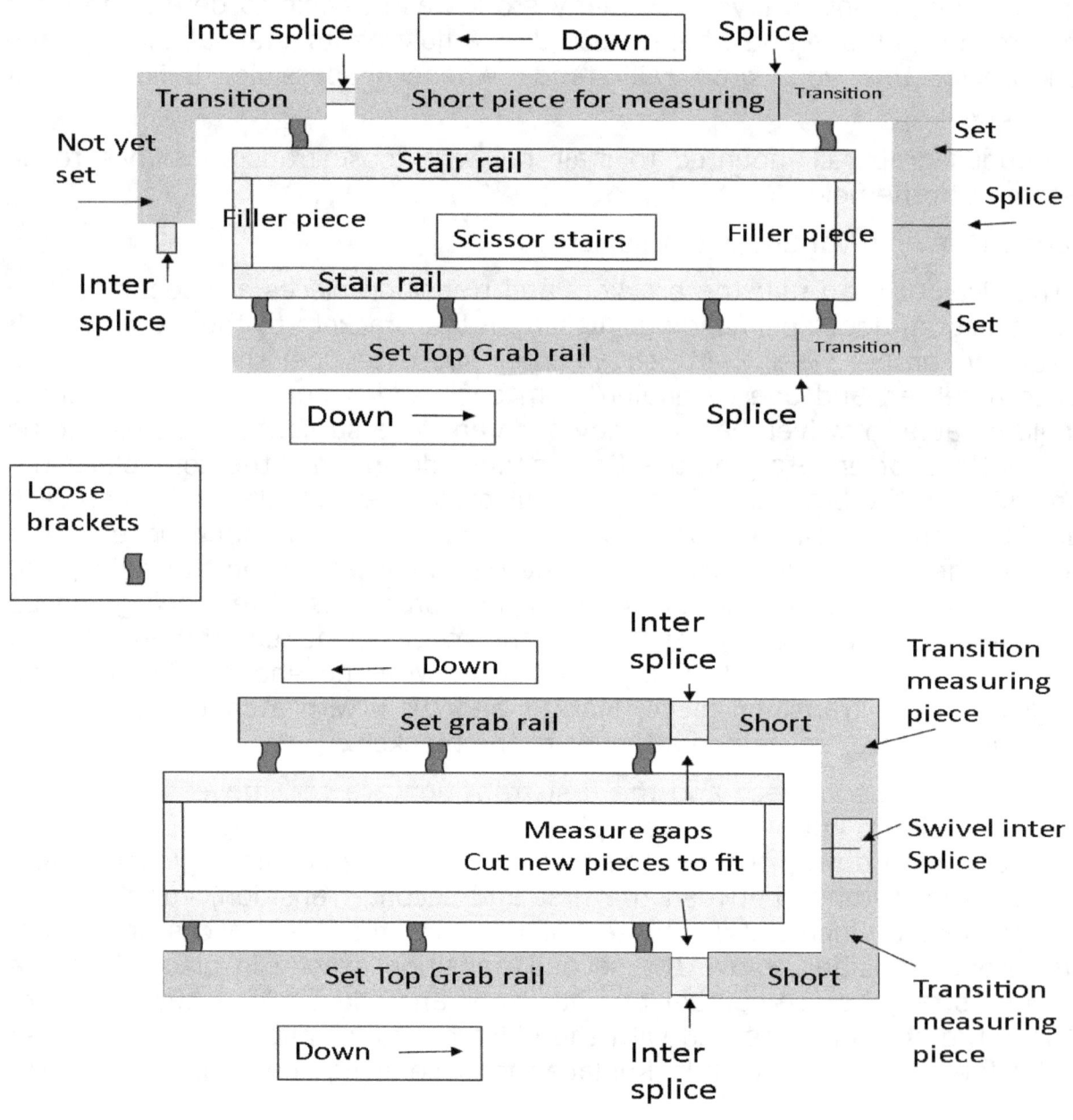

Chapter 8
Installing Yard Railing

Tools:

1. Posthole digger

2. 100' metal tape

3. Stakes

4. 1 x 2 furring strips (for props)

5. Quick grip and 4" swivel jaw C clamps

6. Ratchet straps

7. Shell bar

8. String line

Run a string, just inside the property line. Drive the stakes a few feet before the rail begins and a few feet past where it ends. Tighten the string so when the railing is set there will be about 1/4" between the string and the outside of the posts. Layout the post spacing with the 100ft tape, mark the centerline of the post with a cross—spray painted on the ground. Hold the string out of the way with your leg while digging, to avoid cutting the string with the posthole diggers. Dig the holes a few inches deeper than the post, a shell bar will be necessary in hard ground, a chipping hammer or tractor with an auger might be necessary when working in coral. Place the first section of railing in the holes, place a furring strip prop on each side of the first post in the form of an x and clamp. This will keep the section upright while connecting the rest of the run, more double propping might be necessary, depending on the length of the run.

Place the first post 1/4" inside the string and plumb. Fill the hole with dry concrete and pack it tightly with a rod or piece of furring strip. Continue the process until the run is complete. Pick up the low spots and repack until the run is level. Wet the concrete thoroughly, it will harden overnight.

Gates

Position the gate posts level, plumb and square, with the opening less than 2" larger than the gate. Hold each post in place with a double prop. Mix concrete and anchoring cement in a 5gal bucket. Pour it in the holes — it will dry in 15 —20 minutes. Mount the hinges to the gate, clamp the hinges to the gatepost with C clamps. When the gate is evenly spaced and operating properly, screw the hinges to the post and mount the latch.

Chapter 9
Glass Railing

A. Installing Glass Panel Railing

First, install the aluminum railing. Make sure the flat rubber track is on the outside, and the wedge rubber track is on the inside.

The first choice is for the flat rubber to be installed at the shop.

Second choice is to install the flat rubber on the ground during down time.

Third choice is to install the flat rubber after the rail is installed, just before the glass panel is installed. Once the flat rubber is installed, place a rubber block two or three inches from the end of each panel.

Panel placement: with two people working as a team, slide each panel up into the top cap until it clears the bottom channel, then lower the panel down into the bottom channel until it rests on the rubber blocks. When all the panels are in place, adjust the space between the panels and roll the wedge rubber in the track. Soapy water spray will lubricate the wedge rubber, if the slot is too tight for the wedge rubber. Cut long thin wooded wedges and hammer them in the slot against the glass, compressing the flat rubber and opening the slot a little more. Then proceed to install the wedge rubber.

B. Installing Structural Glass

Structural glass is normally set in a heavy aluminum base shoe or a long stainless steel set in a slot on a curb of parapet wall.

The base shoe is mechanically fastened; the actual glass is set the same way.

This starts with the contractor creating a slot when he pours the slab or wall. A stainless steel track is set in the slot. Hopefully, the contractor will set the track when he makes the pour. If you have to do it, find out if the track is to be set higher to allow for stucco, and how much.

Plug the ends of the track with Styrofoam, insulations, or duct tape, to keep the anchoring cement out while pouring. Shim the track from underneath for height. Add weight to hold down any bows in the track, and shim the sides to take out any twist. Keep the track level both ways or it will show later.

Setting the glass

Place rubber blocks in the track one near the end of each pane of glass. Place the panes on top of the rubber blocks, they can be leaned in or out without falling. Be very careful not to let the edge of the pane, come in contact with concrete, metal, or another pane of glass. It will break or chip. Use wood for wedges and spacers, plastic for shims and spacers.

Once all the panes in the run are in the track, adjust the space between panes, using spacers to maintain equal spaces. Stand up the first piece; it should be level if the track is level. It can be shimmed up or dropped down a little, if necessary. If the plans call for an aluminum cap installed on the glass, the cap will hide slight variations in height between panes. Place the level on the edge of the glass, and raise or lower it until it is plumb, keeping the space between the panes equal vertically. Once the pane is plumb both ways, wedge it in like a post. When the run is complete, install the top rubber and carefully install the cap with a dead blow hammer. The cap will hold all the panes in alignment with each other.

If the planes do not call for a cap, the right height is the priority. Stand up the first piece, level, plumb, and wedge it. Continue the process until the run is complete and the height of each pane is consistent. Build clamps to keep the panes in alignment. Use two pieces of 3/4" wood approximately 2" wide and 4" long, drill a 1/4" hole through the center of both pieces. Connect them together with a 1/4" through—bolt and wing nut. With the center of the bolt in the space between the panes and a piece of wood on each side, slide the clamp down until the top of the panes are visible. Tighten the wing nut. A sliver of clear tubing, smaller then the thickness of the glass, can be slipped over the bolt — between the woods to keep the bolt from touching the edge of the glass; as a safety precaution.

C. Anchoring Structural Glass

Mix the anchoring cement to a little thinner consistency than usual. Use a Rubbermaid brand dust pan to easily funnel the anchoring cement into the slot, between the glass and the track. The anchoring cement must flow under the glass and fill both sides to the top. Be sure to mix extra to accomplish this in a single pour. The glass may be too tall to reach over to the other side. If the run is long or conditions are windy, it may be necessary to anchor one pane at a time. Build a dam between panes, out of Styrofoam or insulation. Sand works well and can be blown out with a vacuum. If the glass is low enough to reach over, The anchorin

Chapter 10
Mixing, Pouring, Anchoring Cement or a/c

A. Tools and supplies

1. A five gallon bucket to mix in.

2. A bucket of clean water.

3. A rinse water bucket for tools, sponge, and hands.

4. A sponge for cleanup and quickly removing rainwater from the holes before the railing is put in.

5. A turkey baster, can be used to remove rain water from the holes, after the railing is in, if there is enough room alongside the post. (slow process, only to be used if power is an issue, or just a few holes to do)

6. A small wet vacuum, with a piece of clear tubing, taped to the vacuum hose that will flex alongside the post and suck the rain water out of the hole.

7. A container for measuring water. (1qt. or larger)

8. A large stirring spoon for scooping dry a/c and scooping a/c too thick to pour out of containers.

9. A wire egg whisk. Insert the wire handle into a piece of electrical conduit or copper tubing; crush the tubing, extending the handle. (ideal for mixing small amounts up to 7 or 8 holes without power)

10. A medium paint paddle and electric drill, for mixing large quantities.

11. An empty 2— Liter plastic bottle with a cone shaped neck to funnel the a/c into the holes. Cut the top off just below the ring. Next mark from the top about 1/3 of the way down, mark across the face— 1/3 of the way around, then back up to the top 1/3 the way around from the first mark. Cut it out — there should be a 1/3 section missing from the one side of the top of the bottle. Once filled, it should fill two —three holes.

12. An empty anti—freeze container, for pouring one handed into the 2—Liter bottle funnel. Cut the top off the anti—freeze container leaving about 2" on the handle side, to enable carrying with just a finger or two. This container can be easily filled from a 5gal. Bucket.

13. A 1gal jug with a cone shaped neck or a 1gal Clorox bleach bottle. Cut using the 1/3 formula. This can pour about 5 holes and can be filled from the 5gal bucket.

Two liter funnel

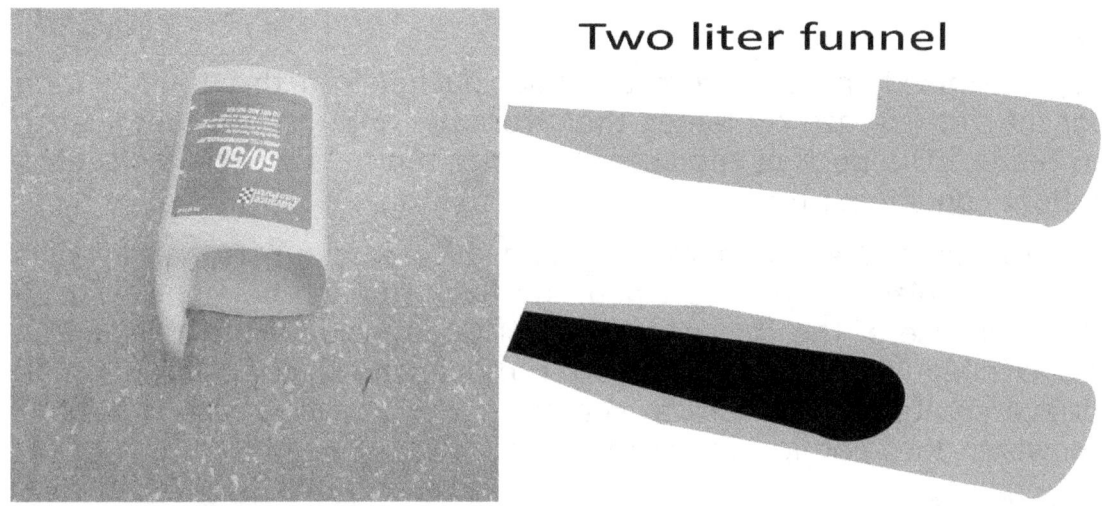

B. Mixing anchoring cement or a/c

Pour 1qt clean water into a 5gal bucket. (dirty water quickens a/c drying time). Add a/c and stir with modified egg whisk, keep adding a/c and stir until thick pea soup consistency. When you pull the wire whisk out of the mix, the a/c should run off. If it drips off it's too thin, add more mix. See how many holes this amount will fill. Increase the water and a/c amounts until the mix fills the required amount of holes. Once the batch mix is determined, find a container that fits in an empty 5gal bucket. Pour in the proper amount a/c powder, for the desired amount of holes; mark the level inside the container. Cut the top of the container off at the mark. Affix a tie wire handle to the container. Set the container inside a bucket, pour the a/c powder in, overflowing the container, shake off the excess. Pour 80% of the powder in the pre—measured water and mix; add the other 20% while mixing. The mix might need a spoonful of a/c or a handful of water to bring it to the right consistency. a/c can be mixed in under two minutes, using unskilled labor.

C. Cold weather mixing

a/c takes 15—20 minutes to harden in 70—degree weather.

a/c could take hours to harden in 40—degree weather. Use hot water to speed up the process. Use a 30—40 cup coffee maker to heat water and mix it with about 3gal of water in a 5gal bucket.

D. Mixing Old or over heated a/c powder

Sometimes a/c can be old or over heated as a powder. This speeds up the hardening process to as little as 30 seconds. Add ice to the water to slow down the process.

E. Pouring a/c

Once the a/c is mixed, pour it from the 5gal bucket into the 1gal pouring container or the anti—freeze container, then from the anti—freeze container to the custom funnel. The neck of the custom funnel can be squeezed to control the flow in tight conditions; mess free.

Fill the hole to about 3/4" from the top and allow the a/c to settle while pouring the other holes. On the next pour, go back to each hole and over fill them a little. Because of the thickness, the surface tension will the allow the hole to be over filled without letting the a/c run over the side. For holes on an uneven surface create a sand dam to contain a/c. This will create a positive drain so rainwater does not lie against the post.

F. Anchoring cement

There are several types of a/c; all with different chara/cteristics.

1. Railing requires quick setting.

2. It must be iron free

3. It must be non—shrink

4. It must not deteriorate

5. It must be aluminum friendly

6. It must mix thoroughly and not separate to the bottom of the bucket

If it contains Jipson — it will deteriorate

Bonsal meets all requirements and can be ordered in 50lb. buckets at Home Depot; They might also drop ship to the jobsite on 24 bucket pallets.

You can also use two—part epoxy products; they are expensive, difficult to mix, messy with an overnight drying time and not practical for standard rail setting operations.

Chapter 11
Anchoring in Concrete

A. Dry Drilling Concrete

Use a hammer drill for drilling in concrete block or soft concrete with no aggregate in the mix. If the masonry bit burns up by the second hole, the aggregate is too hard to drill through with a hammer drill.

Rotary hammers have a piston action and drills through all types of masonry quickly with little pushing effort. They use SDS quick disconnect drill bits.

Drilling accurately in concrete can be difficult; the vibration of the drill bounces the bit off the mark. First, mark the hole with a cross, start drilling after the bit bounces off the mark, drill in about 1/16" and angle the drill toward the mark. The drill bit will slowly drift. When it's on the mark exactly, straighten out the drill and drill the hole.

If you hit rebar, angle the drill until it drills alongside the rebar and finish drilling the hole.

B. Wedge Anchors

A wedge anchor is threaded on one end; the other end is the same size but tapers down as it goes toward the threads. A short split sleeve sits in the taper. There is a nut on the threaded end. The anchor is hammered into the hole. As the nut tightens, the split sleeve rides the taper and expands; wedging between the concrete and the end of the anchor. It will not pull out.

When selecting a wedge anchor, make sure the threaded end is chamfered. If the threads go all the way to the end, hammering the anchor in can easily mushroom the threads, and the nut cannot be taken off. Many times the nut jams and cannot be tightened.

Be sure the anchor has threads down to the taper. If there is an air bubble or the hole is a little shallow, you will not run out of threads while tightening the nut. (Any excess can be cut off with a 4 1/2" grinder.)

The sleeve has three large barbs or two small humps. The three large barbs catch better, but are harder to drive in and there is more chance of mushrooming the threads. Two small humps have a tendency to slip on anchors less than 3/8" in diameter. Neither will hold in lightweight concrete.

Tips

When driving in the anchor, the sound will let you know if it bottoms out. Do not hit it again, it will only jam the nut or damage the anchor.

If the anchor is not chamfered, damage can be avoided by placing a 2x4 between the anchor and the hammer when driving the anchor in. The wood will not damage the threads.

Use an electric impact wrench for tightening or it is a slow process.

C. Sleeve Anchors

There are two types:

Type #1 is a threaded cone; a threaded rod runs through the cone, with a nut and washer on the other end of the threaded rod. A long split sleeve runs from under the washer and nut to just over the beginning of the cone. As the nut tightens, the cone pulls into the sleeve until the sleeve is wedged against the concrete.

They work in softer concrete, block, brick, or applications through multi materials, roofs, and some floors.

Type #2 is a bolt head with a hollow threaded shaft accepting a threaded rod with a built in cone on the other end; a sleeve runs from the bottom of the threaded hollow shaft to just over the small end of the cone. When tightened, the threaded rod is pulled into the shaft, the cone end wedges the sleeve against the concrete. The bolt head is a much cleaner look then a nut.

Caution the length of the anchor is crucial for the application. If the threaded rod bottoms out in the threaded hollow shaft before the cone end wedges the sleeve, the anchor will not tighten.

D. Tapcons

Drill a 3/16" hole for a 1/4" Tapcon, they go in fast, and can be easily removed. They can be used in concrete, block, and brick, even wood. Great for grab rail, and plate mount. Note: use a 1/4" stain steel washer against aluminum.

Tips:

Have several 3/16" plastic shields with you when using 1/4" Tapcons. If one spins, pull it out, set the screw gun in reverse, and push the Tapcon into the hole in reverse. This will enlarge the 3/16" hole to 1/4". Push in a 3/16" plastic shield, set the screw gun back to forward and screw the Tapcon back in. It will hold.

Stainless steel Tapcons spin easier than steel Tapcons. To remedy this, Tapcon now makes a slightly larger Tapcon that uses a 7/32nd bit.

E. Ramset Redhead 3/8"

These are the best 3/8" anchors I have found.

They work like a Tapcon; however, for a 3/8" anchor — a 5/16 bit is used. Some use a 3/8" bit. The bolt actually threads the concrete so same anchor can be removed and replaced several times — using the same hole.

Most other anchors require re—drilling and shimming with plastic or wire.

F. Lags Screws And Shields

For 1/4" lag screws, use a 3/16" plastic shield in a 1/4" hole. They work well for grab rail and plate mount railing. They are removable and easy to find in stainless steel. Use 1 1/2" long in block walls, drill a 1/4" hole, then attach a shield on the end of the lag with a twist. Hammer the lag so the shield is completely in the hole. Screw in the lag.

When going into concrete walls with a 2" or longer lag, drive the shield in further; because the threads are about 1 1/4" long no matter what the lag screw is.

Lag screws 5/16" and up will use lead shields.

G. Pin Grips

Pin grips have a pan head on one end of a shaft. The other end is split in fourths about 3/8" up the shaft. A rod protrudes about 1/4" out of the pan head, traveling inside the shaft and stops at the top of the split on the other end. When the shaft is driven into the pan head, the other end opens and wedges against the concrete. They do not come out, and are considered permanent.

Tips

The drive is so short; the risk of loosening is high. Air bubbles or other defects in the concrete are common. Use only for decorative applications or for things that won't be handled.

H. Drive Nails

Drive nails have a pan head on one end of the shaft, the other end is split in half about 1" up the shaft. A nail protrudes about 1" out of the pan head, traveling inside the shaft, stopping at the top of the split at the other end. When the nail is driven into the pan head, the other end opens and wedges against the concrete. They do not come out and are considered permanent.

Tips

They hold better than pin grips, but the downside is the nail is hard to hit in tight spots. Get a long 5/8" carriage bolt; thread the nut leaving the threads approximately 1/16" down, causing a depression that will keep from slipping off the nail while setting the drive nail.

I. Crook Pins

Crook pins have a pan head on one end of a shaft and a crook in the shaft near the other end. When driven into the hole, the crook wedges against the concrete.

Tips

They hold very well, especially the larger sizes. They also come in plastic for lighter duty applications.

J. Epoxy Anchoring

Epoxy anchors are usually cut threaded rods. Drill the hole, apply epoxy, set rod and wait. Epoxy is usually used for plate—mount applications and has a long drying time. If possible, use Tapcons through fender washers in enough locations to secure the railing, and then use epoxy anchor rods to fill the rest of the holes. When the epoxy dries, tighten the nut and replace the Tapcons with epoxy rods.

Important tips: do not drill an oversize hole; drill a 1/2" hole for a 1/2" rod. Most two—part epoxy comes in a double tube with a mixing chamber. Do not use the first inch out of the mixing chamber, it does not mix properly and will not harden.